Cargo Hose Rupture and Release of Anhydrous Ammonia During Offloading of a Werner Transportation Services Cargo Tank Motor Vehicle at the Tanner Industries Plant Swansea, South Carolina July 15, 2009

Accident Summary Report

NTSB/HZM-12/01/SUM
PB2012-917002

National Transportation Safety Board

NTSB/HZM-12/01/SUM
PB2012-917002
Notation 8391
Adopted April 12, 2012

Hazardous Materials Accident Summary Report

Cargo Hose Rupture and Release of Anhydrous Ammonia
During Offloading of a Werner Transportation Services Cargo
Tank Motor Vehicle at the Tanner Industries Plant
Swansea, South Carolina
July 15, 2009

**National
Transportation
Safety Board**

490 L'Enfant Plaza, S.W.
Washington, D.C. 20594

National Transportation Safety Board. 2012. *Cargo Hose Rupture and Release of Anhydrous Ammonia During Offloading of a Werner Transportation Services Cargo Tank Motor Vehicle at the Tanner Industries Plant, Swansea, South Carolina, July 15, 2009.* **Hazardous Materials Accident Summary Report NTSB/HZM-12/01/SUM. Washington, D.C.**

Abstract: On July 15, 2009, about 8:00 a.m., a cargo transfer hose ruptured shortly after transfer of anhydrous ammonia began from a Werner Transportation Services, Inc. cargo tank truck to a storage tank at the Tanner Industries, Inc. facility in Swansea, South Carolina. A white cloud of anhydrous ammonia, a toxic-by-inhalation gas, moved from the parking lot of the facility across U.S. Highway 321 to a largely wooded area, where it eventually dissipated. About the same time, a motorist traveling north on the highway drove into the ammonia cloud, apparently tried to get away from the cloud, then got out of her car and died of ammonia poisoning. Seven people went to the Lexington Medical Center emergency department complaining of respiratory problems and dizziness; all seven patients were treated and released the same day. The anhydrous ammonia cloud caused temporary discoloration of vegetation in the area, including the leaves on the trees. Residents in the area sheltered in place, and U.S. Highway 321 was closed until about 2:00 p.m. on the day of the accident. The Lexington County Fire Service arrived on scene about 8:07 a.m. Property damage and losses were limited to the ruptured hose and about 6,895 pounds of the anhydrous ammonia that was released.

The National Transportation Safety Board (NTSB) is an independent Federal agency dedicated to promoting aviation, railroad, highway, marine, pipeline, and hazardous materials safety. Established in 1967, the agency is mandated by Congress through the Independent Safety Board Act of 1974 to investigate transportation accidents, determine the probable causes of the accidents, issue safety recommendations, study transportation safety issues, and evaluate the safety effectiveness of government agencies involved in transportation. The NTSB makes public its actions and decisions through accident reports, safety studies, special investigation reports, safety recommendations, and statistical reviews.

Recent publications are available in their entirety on the Internet at <http://www.ntsb.gov>. Other information about available publications also may be obtained from the website or by contacting:

National Transportation Safety Board
Records Management Division, CIO-40
490 L'Enfant Plaza, SW
Washington, D.C. 20594
(800) 877-6799 or (202) 314-6551

NTSB publications may be purchased, by individual copy or by subscription, from the National Technical Information Service. To purchase this publication, order report number PB2012-917002 from:

National Technical Information Service
5285 Port Royal Road
Springfield, Virginia 22161
(800) 553-6847 or (703) 605-6000

The Independent Safety Board Act, as codified at 49 U.S.C. Section 1154(b), precludes the admission into evidence or use of NTSB reports related to an incident or accident in a civil action for damages resulting from a matter mentioned in the report.

Contents

Figures

Acronyms and Abbreviations

CFR	*Code of Federal Regulations*
DHEC	South Carolina Department of Health and Environmental Control
DOT	U.S. Department of Transportation
EMS	Emergency medical services
FMCSA	Federal Motor Carrier Safety Administration
LPG	Liquefied petroleum gas
NTSB	National Transportation Safety Board
PET	Polyethylene terephthalate
PHMSA	Pipeline and Hazardous Materials Safety Administration
psig	Pounds per square inch, gauge
Tanner	Tanner Industries, Inc.
UL	Underwriters Laboratories Incorporated
Werner	Werner Transportation Services, Inc.

Executive Summary

On July 15, 2009, about 8:00 a.m., a cargo transfer hose ruptured shortly after transfer of anhydrous ammonia began from a Werner Transportation Services, Inc. cargo tank truck to a storage tank at the Tanner Industries, Inc. facility in Swansea, South Carolina. A white cloud of anhydrous ammonia, a toxic-by-inhalation gas, moved from the parking lot of the facility across U.S. Highway 321 to a largely wooded area, where it eventually dissipated. About the same time, a motorist traveling north on the highway drove into the ammonia cloud, apparently tried to get away from the cloud, then got out of her car and died of ammonia poisoning. Seven people went to the Lexington Medical Center emergency department complaining of respiratory problems and dizziness; all seven patients were treated and released the same day. The anhydrous ammonia cloud caused temporary discoloration of vegetation in the area, including the leaves on the trees. Residents in the area sheltered in place, and U.S. Highway 321 was closed until about 2:00 p.m. on the day of the accident. The Lexington County Fire Service arrived on scene about 8:07 a.m. Property damage and losses were limited to the ruptured hose and about 6,895 pounds of the anhydrous ammonia that was released.

The National Transportation Safety Board determines that the probable cause of the accident was Werner Transportation Services, Inc.'s use of a cargo hose assembly that was not chemically compatible with anhydrous ammonia. Contributing to the accident was the lack of explicit requirements by the Pipeline and Hazardous Materials Safety Administration that the motor carrier and the facility carrier verify that the cargo hose assembly is chemically compatible with the product to be transferred before transfer operations begin.

The following safety issues were identified in this accident:

- Use of chemically incompatible hose assembly.

- Inadequate unloading procedure.

- Inadequate cargo hose requirements for passive emergency discharge, monthly inspection, and annual leakage tests.

As a result of its investigation of this accident, the National Transportation Safety Board makes safety recommendations to the Federal Motor Carrier Safety Administration and the Pipeline and Hazardous Materials Safety Administration.

1. The Accident

On July 15, 2009, about 8:00 a.m., a cargo transfer hose ruptured shortly after transfer of anhydrous ammonia began from a Werner Transportation Services, Inc. (Werner) cargo tank truck to a storage tank at the Tanner Industries, Inc. (Tanner) facility in Swansea, South Carolina. A white cloud of anhydrous ammonia, a toxic-by-inhalation gas,[1] moved from the parking lot of the facility across U.S. Highway 321 to a largely wooded area, where it eventually dissipated. About the same time, a motorist traveling north on the highway drove into the ammonia cloud, apparently tried to get away from the cloud, then got out of her car and died of ammonia poisoning. Fourteen people reported experiencing minor respiratory problems or dizziness as a result of the anhydrous ammonia release and were evaluated by emergency medical services (EMS) on scene. Of those 14, 7 displayed symptoms that required EMS to transport them for further evaluation at an emergency department; they were treated and released the same day.

1.1 Accident Narrative

A Werner cargo tank truck, unit No. 3002, a U.S. Department of Transportation (DOT) specification MC 331,[2] arrived at the CF Industries terminal[3] in Tampa, Florida, about 9:39 p.m. on July 14, 2009. The truck was loaded with 39,516 pounds (19.758 tons) of anhydrous ammonia. The truck left the terminal about 10:39 p.m. to travel to Swansea, South Carolina, to deliver the product.

Shortly after 7:00 a.m. on July 15, 2009, the Werner cargo tank truck arrived at the Tanner facility in Swansea, South Carolina. Upon arrival, the driver and a driver trainee checked in at the office, where they were told to park their truck next to the piping manifold near a designated storage tank. About 7:40 a.m., the driver parked the truck, picked up the vapor hose assembly that was lying on the ground next to the manifold, and connected the hose assembly to the vent line on the cargo tank. He then removed the cargo transfer hose assembly from the storage tube on the trailer, checked the pressure valves on the tank, and connected the hose assembly to the liquid discharge fitting on the tank. A Tanner plant employee fastened the other end of the transfer hose assembly to the facility piping manifold. The driver then engaged the power take-off unit, turned on the pump, and checked the tank volume gauge on the side of the cargo tank to make sure product was flowing from the cargo tank to the storage tank. The driver watched the reading on the cargo tank volume gauge drop from 71 percent full to 64 percent full, at which point he told the trainee to oversee the unloading while he completed paperwork in the tractor. The trainee watched the gauge to make sure the unloading was proceeding correctly. The last gauge reading observed by the trainee before the accident indicated that the cargo tank was

---

[1] The U.S. Department of Transportation classifies anhydrous ammonia as a Hazard Class 2 nonflammable gas. Anhydrous ammonia is a colorless liquid or gas that is both poisonous and corrosive and that has an intense, pungent, suffocating odor.

[2] Specification MC 331 is defined in Title 49 *Code of Federal Regulations* (CFR) 178.337 as a cargo tank motor vehicle primarily for transportation of liquefied compressed gases as defined in 49 CFR part 173, subpart G.

[3] Founded in the 1940s, CF Industries was originally known as Central Farmers Fertilizer Company.

about 53 percent full. The trainee later estimated that at that time, 1,500 gallons had been transferred.

Just before 8:00 a.m., about 7 or 8 minutes after the transfer had begun, the trainee heard a pop sound followed by a loud rush of gas. He looked underneath the truck and saw a rupture in the transfer hose assembly directly in front of him. He stated that he saw anhydrous ammonia pluming upward from the rupture, forming a dense white cloud. The trainee immediately pushed the emergency shutdown button on the rear of the cargo tank. Upon doing this, he noticed that movement of the ammonia cloud quickly shifted in the direction of highway 321. He then turned to the facility manifold to find a shutdown switch, but he did not see one because the white ammonia cloud was surrounding the manifold. He then evacuated through the south gate and ran west into a wooded area. Eventually he heard the hose stop and saw the cloud lift soon after.

The driver stated that he also heard a loud pop about 8:00 a.m. and saw a cloud of gas outside the driver's side window. He then turned off the engine to stop the trailer pump; put on his half-face respirator; got out of the truck on the passenger side; and escaped through the white cloud in front of the building along the fence, exited through a gate, and followed the railroad tracks.

1.2 Emergency Response

At 7:57 a.m., the Lexington County Communications Center received the initial 911 call from a Tanner employee at the facility reporting an anhydrous ammonia leak. The first Lexington County Fire Service unit was dispatched at 8:02 a.m.; the fire service arrived on scene at 8:07 a.m. About 8:32 a.m., a Tanner representative in Pennsylvania contacted the Lexington County Emergency Management Department and reported that two plant employees and two drivers were on scene and that a hose failure had resulted in the release of 100 to 500 pounds of anhydrous ammonia. The Tanner representative also recommended to the Emergency Management Department that the community should shelter-in-place as a precautionary measure after the anhydrous ammonia release. Fire service search teams found that most residents had self-evacuated.

Personnel from the Lexington County departments of public safety and emergency management arrived on scene at 8:55 a.m., and a command post was established. At 9:05 a.m., the first unified command meeting took place; the meeting included representatives from the Lexington County Emergency Management Department, fire services, EMS, and the sheriff's office.

At 9:12 a.m., fire services began sending search teams into the populated areas adjacent to the plant and to U.S. Highway 321 that had been in the path of the plume. They found two people who complained of respiratory problems who were transported for medical treatment. About 9:25 a.m., the search team discovered the fatally injured victim located outside of her vehicle on the side of the highway and notified EMS immediately. By 10:30 a.m., search teams had checked several homes but found no one at home.

The South Carolina Department of Health and Environmental Control (DHEC) arrived on scene at 9:50 a.m. and began taking air samples; by 11:25 a.m., the DHEC reported "all clear"[4] on the air sample readings and declared the area safe for the public.

During the second unified command meeting at 10:35 a.m., it was reported that about 1,785 gallons of anhydrous ammonia had been lost from the ruptured transfer hose assembly and that emergency workers were continuing to search for victims in the area. At the third unified command meeting, at 11:40 a.m., it was reported that all people had evacuated from the scene, that the DHEC had given the "all clear," and that the Tanner plant was safe. Three additional meetings took place at 12:35, 1:53, and 2:30 p.m. At the final unified command meeting, emergency response activities were reviewed, and all emergency service parties involved agreed that the scene was safe. The command post was closed at 2:50 p.m., and all units were cleared from the scene by 4:33 p.m.

1.3 Injuries

At the time of the accident, about 8:00 a.m. on July 15, 2009, a woman traveling northbound on U.S. Highway 321 past the Tanner plant drove her car into the dense cloud of anhydrous ammonia covering the roadway and was fatally injured. In an apparent attempt to escape the ammonia cloud, the motorist had driven her car off the road and then got out of the car. She was found about 9:25 a.m. on the ground about 10 feet from her car. According to the Lexington County Coroner's Office, the cause of the driver's death was ammonia poisoning.

A total of 14 people reported minor respiratory problems or dizziness as a result of the anhydrous ammonia release; each person was evaluated by EMS on scene. Seven of these people displayed symptoms that required EMS to transport them to the emergency department at Lexington Medical Center for further evaluation. All seven patients were treated and released the same day. The remaining seven did not display symptoms that required them to be transported by EMS to the Medical Center.

---

[4] *All clear* in this instance means that no trace amount of anhydrous ammonia was detected in the air sample readings.

2. Investigation and Analysis

2.1 Facility and Carrier Information

Tanner, of Southampton, Pennsylvania, owned and operated the Swansea, South Carolina, facility at which this accident occurred. Tanner is a family-owned and -operated company established in 1954. The company also was registered with the Federal Motor Carrier Safety Administration (FMCSA) and the Pipeline and Hazardous Materials Safety Administration (PHMSA) as a private carrier, transporter, and shipper of various hazardous materials, including anhydrous ammonia.

The Tanner facility in Swansea became operational in 1996. The Swansea facility manufactures ammonium hydroxide and distributes ammonium hydroxide and anhydrous ammonia. Tanner purchases anhydrous ammonia from vendors and has the product transported to its Swansea facility by cargo tank truck or rail car. The anhydrous ammonia is then either repackaged for bulk or non-bulk distribution to customers or blended into ammonium hydroxide on site.

In the 10 years before this accident, no unintentional releases of either anhydrous ammonia or ammonium hydroxide that meets the U.S. Environmental Protection Agency's Risk Management Program reporting requirements had occurred at Tanner's Swansea facility. Additionally, PHMSA conducted a security inspection of the Tanner facility on August 28, 2008, and no violations of the Hazardous Materials Regulations were noted.

The cargo tank truck involved in this accident was owned and operated by Werner, of Gainesville, Georgia. The company's business consists primarily of bulk transportation of anhydrous ammonia to its customers. Werner also transports flammable gases, including liquefied petroleum gas (LPG) and butane. At the time of the accident, all 21 of Werner's hazardous materials cargo trailers were DOT specification MC 331 cargo tanks that are authorized for the transportation of liquefied compressed gases, including both LPG and anhydrous ammonia.

Werner operates primarily in the eastern half of the United States; however, the company also has a loading facility in Ontario, Canada, about 7 miles from the U.S.-Canada border. Most of Werner's business consists of dedicated contract trips from large loading facilities to power plants. Werner does not maintain any fixed facilities other than its principal place of business in Gainesville where most of its administrative business occurs.

2.2 Accident Hose Assembly

The ruptured transfer hose was manufactured by Durodyne, Inc.[5] in 2005. The hose, Durodyne product number DD-G-220, was designed for LPG transfer only and was constructed

---

[5] In August 1999, Durodyne was purchased by Argo Tech Costa Mesa. In 2007, Eaton Corporation purchased Argo Tech, including the Durodyne unit.

of several different layers of material that are chemically compatible with LPG, but not with anhydrous ammonia. The innermost layer of the hose was made of nitrile rubber, followed by three sequential layers of polyethylene terephthalate (PET) yarn braids encased in chloroprene rubber and an outer layer of neoprene rubber. The PET yarn braids are used to mechanically reinforce the hose, and they provide the hose with the majority of its strength. The hose had a nominal inner diameter of 2 inches, an outer diameter of 2 3/4 inches, and a bend radius of 14 inches. It was rated for use between -40° F and 180° F. The hose had been approved by Underwriters Laboratories Incorporated (UL) and met the UL 21[6] standard for LPG hose. The maximum allowable working pressure of the hose assembly was 350 pounds per square inch, gauge (psig), and the minimum specified hydrostatic burst strength was 1750 psig, in accordance with UL 21.

The hose was imprinted with text. One side of the black neoprene cover of the hose featured a blue Mylar stripe extending along the length of the hose with the Durodyne logo and "DD-G-220 LPG TRANSFER ONLY 350 PSI MAX WP" printed in black. The phrases "To prevent serious injury or property damage use for intended purpose only," "Warning: Use of damaged hose could be hazardous; inspect hose and couplings prior to each use," and "Textile reinforcements meet UL21" were also embossed along the imprinted blue Mylar stripe. The opposite side of the hose had "DURODYNE DD-G-220 LPG HOSE UL21 ISSUE E-7874 MH29239 SPEC DD-G-220 TEXTILE BRAID WP 350 PSIG 4Q05 INSPECT HOSE BEFORE USE" embossed on an imprinted stripe extending the length of the hose. (See figure 1.)

Figure 1. Blue Mylar and imprinted stripes on accident hose.

---

[6] UL standard 21, for LPG hose, covers hose and hose assemblies in sizes up to and including a nominal internal diameter of 4 inches for conveying LPG.

Smart-Hose Technologies of Philadelphia, Pennsylvania, purchased the LPG transfer hose from Durodyne and installed the Smart-Hose safety system to fabricate the complete accident hose assembly as a Lifeline III LPG Transfer Hose. The Smart-Hose safety system consists of an internal cable running through the bore of the hose connected to specially designed unseated flapper valves located on each end of the cable. In the event of hose or coupling separation, catastrophic hose rupture, or excessive hose stretching, the system is designed to shut off the flow of LPG in both directions as the flapper valves release and instantly seat. (See figure 2.)

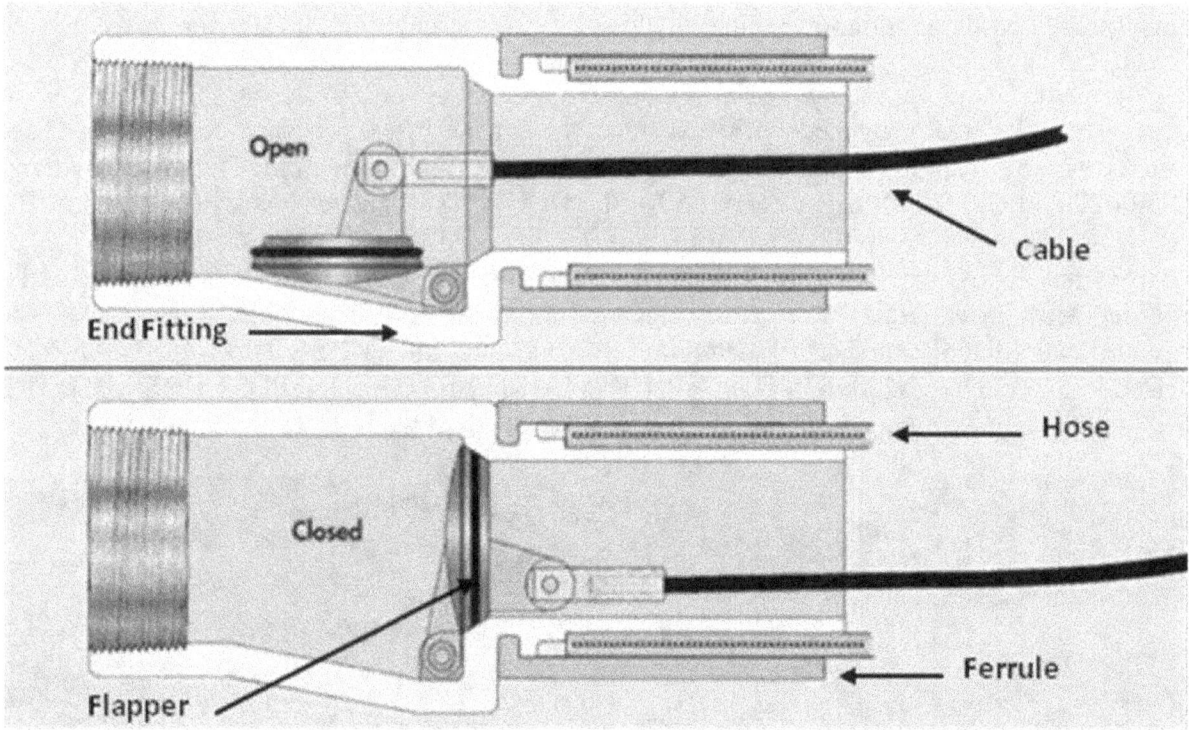

Figure 2. Schematic depicting Smart-Hose safety system.

The Smart-Hose assembly consisted of a 238-inch-long 1/4-inch 1x19 galvanized steel strand cable with a nylon coating down the bore of a 216-inch-long piece of Durodyne LPG transfer hose. Each end of the cable was connected to a valve flapper made from 316 stainless steel.[7] The flappers were connected to 2-inch-diameter 316 stainless steel female national pipe thread end fittings on each end of the hose. The end fittings were secured to the hose by 2-inch ferrules crimped onto each end of the hose, which completed the Smart-Hose assembly and held the cable in compression. Records indicate that the hose assembly was 222 inches (18 1/2 feet) long. Smart-Hose issued a new hose test certification for this hose assembly, serial number 10573, on October 18, 2005.

---

[7] Type 316 stainless steel is an austenitic chromium nickel stainless steel containing molybdenum. Because of its superior corrosion and oxidation resistance, good mechanical properties, and fabricability, 316 stainless steel has applications in many sectors of industry, including its use for the manufacture of tanks and storage vessels for corrosive liquids.

Each end of the hose also had a male acme hammer lock coupling made of cast iron and carbon steel that threaded into each end fitting. This type of coupling is not acceptable in LPG applications because of sparking issues, but it is appropriate for anhydrous ammonia applications. According to Smart-Hose Technologies, these couplings were not installed by or purchased from Smart-Hose at the time Smart-Hose completed the hose assembly. Werner purchased the transfer hose assembly from Gas Equipment Company, Inc. of Indianapolis, Indiana, on December 20, 2005. The invoice for the purchase does not include any information about the couplings, such as whether the couplings were purchased from or installed by Gas Equipment Company. Additionally, no other records or receipts were found that identified the company that installed the couplings. As a result, National Transportation Safety Board (NTSB) investigators were unable to determine when the hammer lock couplings were installed and who installed them.

2.3 Postaccident Testing and Analysis of Accident Hose Assembly

On September 9, 2009, the LPG transfer hose assembly was examined at the NTSB's Materials Laboratory in Washington, D.C., in the presence of the parties to the investigation. The overall length of the hose assembly was 18 1/2 feet. The measured length of the hose from end fitting to end fitting was 18 feet. At the time of construction, Smart-Hose certified the length of the hose assembly, not including the couplings, as 18 1/2 feet.[8] The rupture in the hose was about 5 1/2 inches long (see figure 3). The centerline of the rupture was located 131 inches from the A end[9] and 91 inches from the B end of the hose assembly.

---

[8] Smart-Hose literature states that these hoses may contract up to 3 percent when pressurized. Therefore, an 18 1/2-foot hose could contract 6 inches or more (that is, 3 percent of 18 1/2 feet, or 222 inches, is 6.66 inches).

[9] The "A" and "B" ends of the hose assembly were arbitrarily chosen and labeled by the NTSB's Materials Laboratory for reference use only.

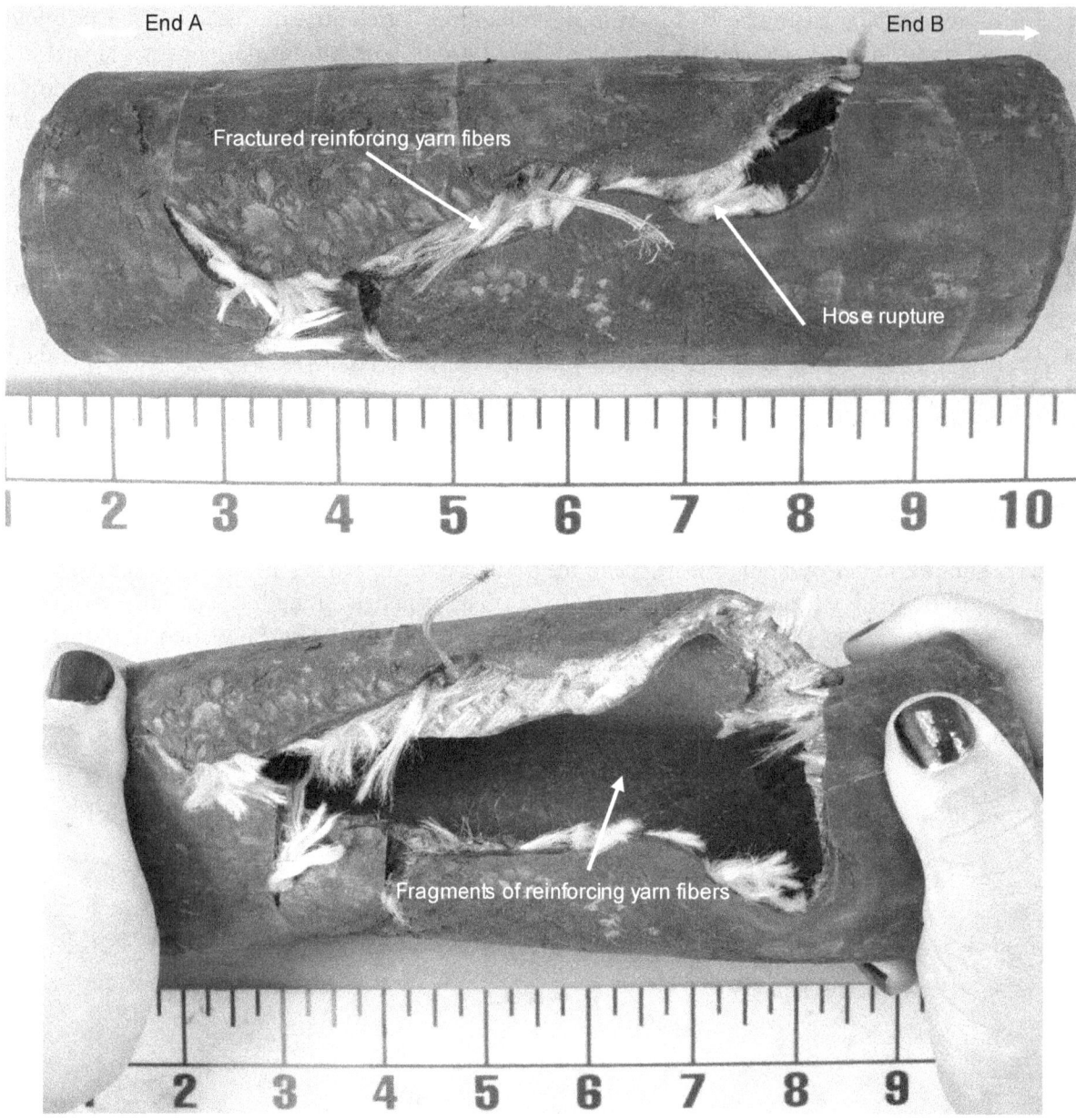

Figure 3. Views of outside (top) and inside (bottom) of ruptured section of accident hose.

The exterior surfaces of the hose assembly contained only superficial abrasions. No gouges, slices, or other defects were noted on the surface of the hose, including the areas adjacent to and abutting the rupture area. The black text on the blue Mylar stripe was abraded in multiple locations so that the stenciling along the length of the hose could not be read; however, several sections of the hose assembly were free of abrasions or stains, and the text could be read. The embossed text on the imprinted line could be read along the length of the hose assembly.

Fractographic[10] evidence indicates that the rupture in the accident hose assembly initiated on the interior wall of the hose and propagated outward. The NTSB's Materials Laboratory identified a definitive fracture origin on the surface of the fracture that was indicative of relatively slow crack growth. (See figure 4.) Several secondary cracks were noted in the interior wall of the hose near the fracture origin; however, the interior surface did not appear to be degraded from anhydrous ammonia exposure.

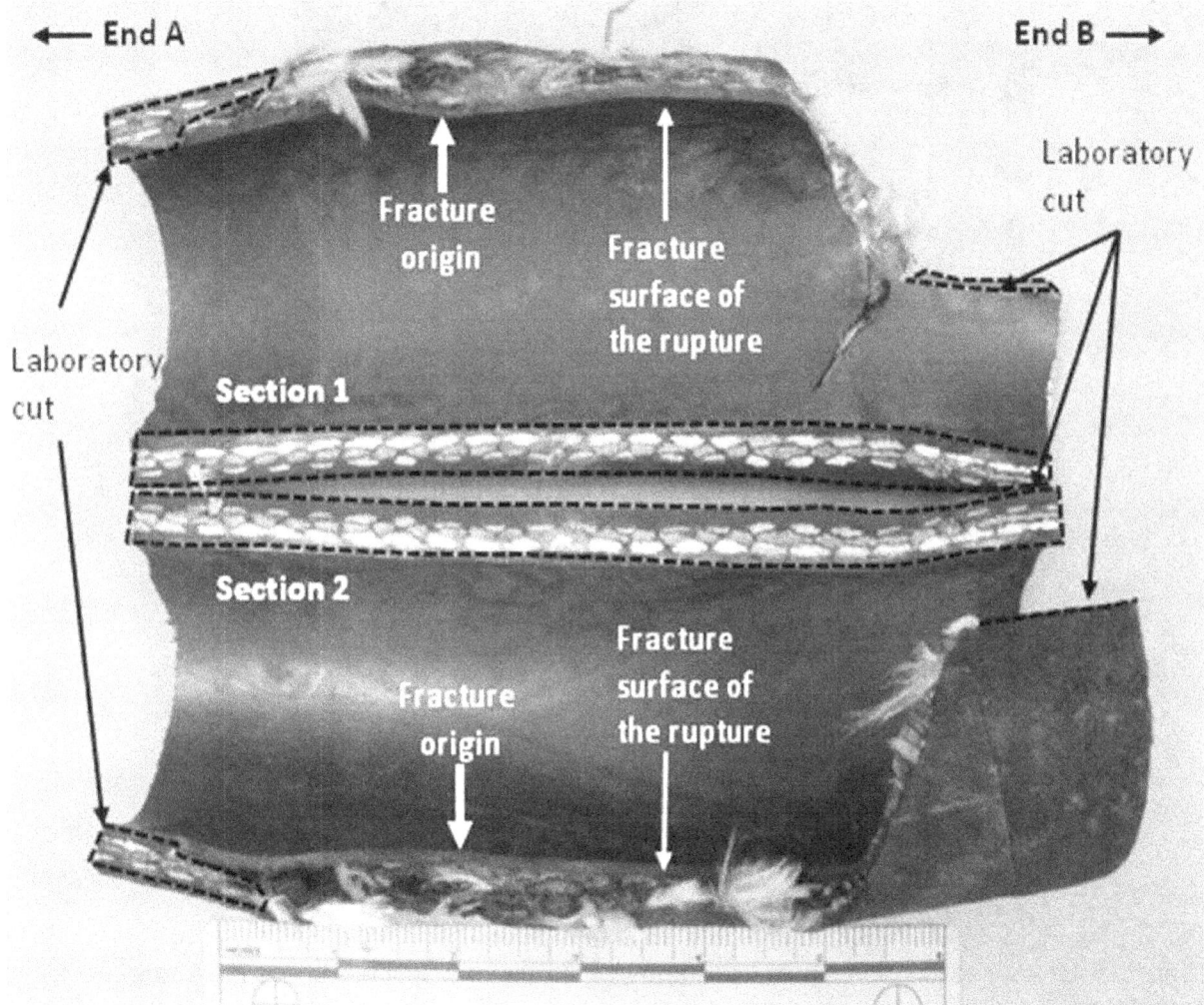

Figure 4. Section of ruptured accident hose cut to reveal fracture surfaces.

The reinforcing fibers in the two innermost PET braid layers of the hose assembly appeared to be severely damaged on both halves of the fracture surface along the entire length of the ruptured area. In some parts of the ruptured area, the fibers were clumped together and appeared to be encased in salt-like particles. Laboratory analysis of the hose assembly revealed that the fibers in the two innermost PET braid layers were degraded to the point that they were brittle and friable when strained or mechanically flexed. (See figure 5.)

---

[10] *Fractography* is the study of the fracture surfaces of materials.

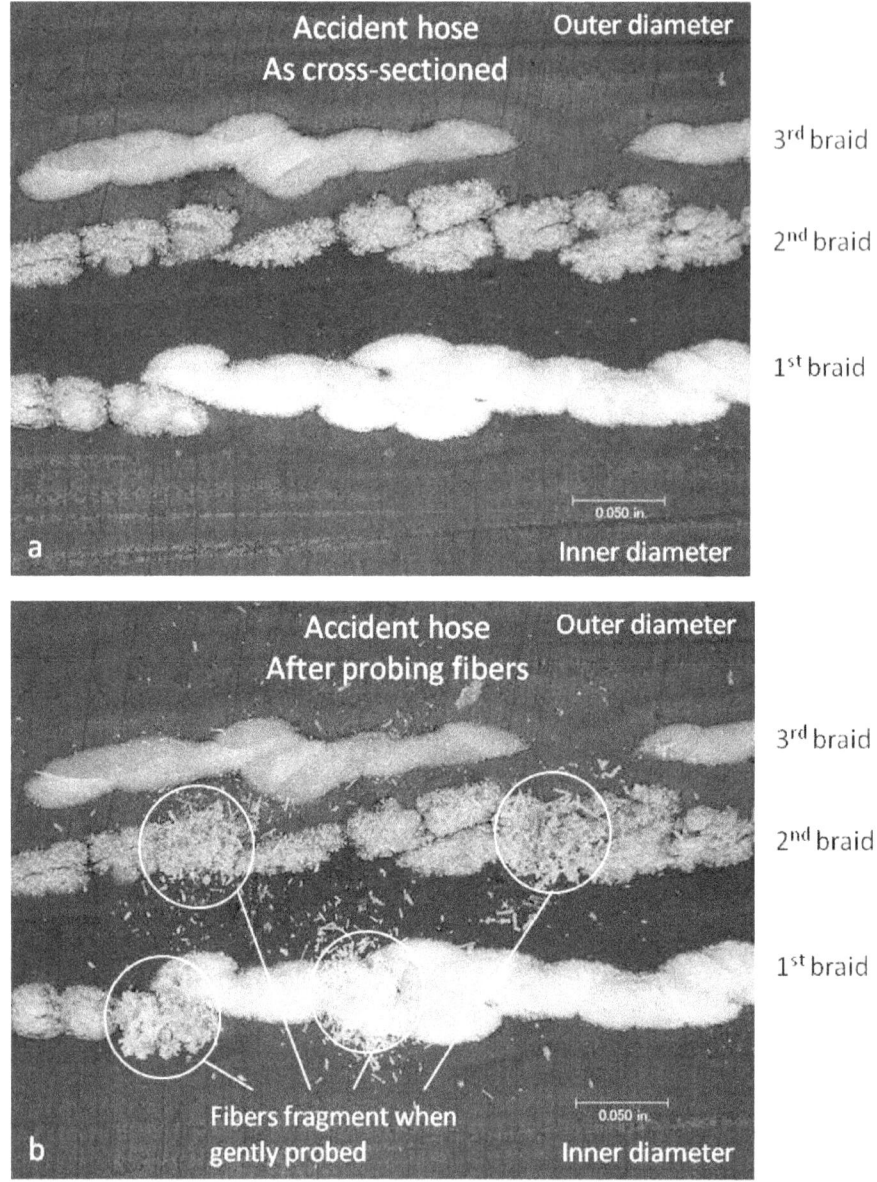

Figure 5. Cross-section of accident hose showing brittleness of fibers in two innermost PET braid layers.

The PET used for the yarn braids is not chemically resistant to anhydrous ammonia or ammonium hydroxide. As documented in several studies,[11] exposure to anhydrous ammonia or ammonia-related compounds results in a chemical reaction (ammonolysis) that can cause PET fibers to degrade and lose strength.

The outer rubber layer of the accident hose had an array of small pinpricks along its length that was intended to allow the product that it was transferring to permeate through the rubber layers of the hose and escape to the atmosphere. The purpose of this is to prevent gas from becoming trapped in the hose wall and damaging the hose; this pinprick design is standard in rubber hoses for LPG and anhydrous ammonia service. When the anhydrous ammonia permeated through the rubber layers of the accident hose, it collected in the interstitial spaces of the fibers within the PET braids. Also, the accident hose was likely exposed to moisture, including humidity and rain, throughout its life cycle. Any absorbed and dissolved moisture contained in the accident hose likely would have converted the trapped anhydrous ammonia to ammonium hydroxide, leading to chemical degradation of the PET fibers.

Testing completed by both the NTSB's Materials Laboratory and an independent laboratory[12] confirms that the PET fibers in the accident hose had sustained chemical degradation that dramatically reduced the strength of the PET fiber. The NTSB concludes that the accident hose failed because it was not chemically compatible with the anhydrous ammonia in the cargo tank that caused the chemical degradation, the loss of mechanical strength, and the ultimate failure of the cargo hose.

---

[11] (a) C. Lorenzetti and others, "Chemical Recovery of Useful Chemicals from Polyester (PET) Waste from Resource Conservation: A Survey of State of the Art," *Journal of Polymers and the Environment*, vol. 14, no. 1 (2006), pp. 89–101. (b) V. Sinha and others, "PET Waste Management by Chemical Recycling: A Review," *Journal of Polymers and the Environment*, vol. 18, no. 1 (2008). (c) M. Khaddaj and others, "Processing of New Materials Using Thermal and Thermo-Vaporous Treatment of Terephthalates," *Journal of Physics*, Conference Series 121 (2008). (d) R. Lamparter and others, "Process for Recovering Terephthalic Acid from Waste Polyethylene Terephthalate, United States Patent 4542239, September 17, 1985. (e) W. Murdoch, "Production of Terephthalic Acid and Ethylene Glycol from Polyethylene Terephthalate by Ammonolysis," United States Patent 6723873, April 20, 2004.

[12] Trace Laboratories also tested the accident hose.

3. Safety Issues

3.1 Use of Chemically Incompatible Hose

When making deliveries, Werner drivers sometimes used facility-owned hoses instead of the hose on the cargo tank vehicle. It is not known how many times the accident hose assembly was used to transfer anhydrous ammonia before it failed.

Following the accident, an NTSB investigator discovered that the accident hose was the LPG transfer hose that was originally assigned to trailer 2322, not the accident trailer (that is, trailer 3002). Further investigation revealed that an anhydrous ammonia transfer hose manufactured by Goodall Canada Inc. was carried on board trailer 2322 at the time of the accident. According to records and statements from Werner, trailers 2322 and 3002 were stored on the same secure lot in Tampa, Florida, for several hours on May 17, 2009. Although Werner stated that neither of its drivers had admitted to exchanging the transfer hoses, no other opportunity existed for the LPG transfer hose to be placed on the accident trailer. Based on its records, Werner estimated that the accident hose was used to unload anhydrous ammonia between 2 and 12 occasions.

The physical properties of hazardous materials vary so greatly that cargo hoses are constructed and intended for use with specific products and cannot be used interchangeably. As previously noted, the hose in the accident hose assembly had internal fibers made of PET, which is not chemically compatible with anhydrous ammonia. Therefore, any cargo hose containing PET fibers would not be suitable for anhydrous ammonia service. The need for chemical compatibility applies not only to the hose material, but to all components of the completed hose assembly, including end fittings and couplers. The accident hose assembly had cast iron and carbon steel couplings that were appropriate for anhydrous ammonia service but inappropriate for LPG service, which requires spark-resistant materials such as brass, bronze, and stainless steel. Therefore, the NTSB concludes that because of the chemical incompatibilities of the hose material with anhydrous ammonia and of the couplers with LPG, Werner's hose assembly was not suitable for use with either anhydrous ammonia or LPG and should not have been carried on the cargo tank motor vehicle that was involved in this accident.

PHMSA's Office of Hazardous Materials Safety issues safety advisory notices to help the public understand significant safety risks. PHMSA's safety advisories are published in the *Federal Register* and provide a description of the safety issue and a recommended action to resolve the issue. The FMCSA, in conjunction with its duties enforcing rules and regulations, conducting inspections, and licensing hazardous materials carriers, also issues safety advisory notices pertinent to cargo tank safety. The two agencies working together could provide the necessary outreach to assist carriers and facility operators in avoiding the hazards associated with the use of chemically incompatible hoses and couplers during loading and unloading operations. Therefore, the NTSB recommends that PHMSA and the FMCSA jointly issue a safety advisory bulletin to inform cargo tank motor vehicle owners and operators, registered inspectors of these vehicles, and transfer facility operators about the circumstances of this accident and actions needed to prevent the occurrence of a similar accident.

Neither the Werner driver who connected the accident hose assembly to the cargo tank nor the Tanner plant employee who connected the hose assembly to the piping manifold recognized that the hose assembly was not approved for unloading anhydrous ammonia. Although the hose was clearly labeled "LPG TRANSFER ONLY," both employees stated during postaccident interviews that they did not make it a point to look at the hose before making the connections to confirm that it was approved for anhydrous ammonia applications.

Werner transport drivers as well as Tanner plant employees were trained in their respective companies' standard operating procedures for loading and unloading. Because Werner transport drivers were required to follow the facility's unloading procedure in lieu of their own, they were given the Tanner unloading procedure for use when making deliveries to Tanner facilities. At the time of the accident, Tanner's anhydrous ammonia unloading procedure directed that hose assemblies be inspected for visual defects prior to beginning unloading operations. The procedure did not direct that the hose assembly to be used be verified as suitable for use with the product being unloaded. Without specific direction, neither the plant employee nor the drivers were required to check whether the accident hose assembly was suitable for anhydrous ammonia.

The Federal regulation covering oversight of the loading and unloading of cargo tank motor vehicles carrying Class 2 hazardous materials,[13] which include anhydrous ammonia and LPG, is found in Title 49 *Code of Federal Regulations* (CFR) 177.840. The regulation does not address the content or structure of private transportation companies' unloading procedures. The regulation includes requirements for (1) the cargo tank motor vehicle carrier (driver) to perform predelivery checks of hose assemblies carried on the vehicle before unloading operations as a means of ensuring that the hose assemblies are free of visual defects and (2) a qualified facility employee to perform the same checks for hose assemblies provided by the facility. The regulation does not specifically require that drivers or facility employees verify that the complete hose assemblies, whether carried on the vehicle or provided by the facility, are suitable for the product that is being unloaded. Further, the regulation does not include requirements for cargo transfer hose assemblies used for other classes of hazardous materials, such as flammable liquids, corrosive materials, and poisons.

Despite the absence of Federal requirements, manufacturers routinely provide guidance to ensure that tanks, containers, piping, and hose assemblies are used for handling and transporting chemically compatible hazardous materials. However, the accident in Swansea demonstrates that although this concept may be understood, adequate steps are not always taken to verify that it is followed. A transfer hose assembly, whether carried on the cargo tank motor vehicle or provided by the facility, may be the single component of a cargo transfer system that is most vulnerable to failure and that potentially has the most severe consequences should a failure occur. When a hose assembly is used to transfer hazardous materials for which it is not chemically resistant, it will eventually fail. Therefore, the NTSB concludes that verification of chemical compatibility of the hazardous material and the transfer hose assembly as part of pretransfer protocol and before loading or unloading begins is critical to minimizing the risk of incompatibility.

---

[13] Class 2 hazardous materials are liquefied compressed flammable, nonflammable, and toxic-by-inhalation gases.

To reduce the likelihood of using a hose assembly that is not chemically resistant to the hazardous material to be loaded into or unloaded from a highway cargo tank, the motor carrier and/or the facility carrier should not only visually inspect the cargo hose assembly for defects, but also verify the chemical products that can be safely transferred through the hose assembly. Verification can be accomplished by noting markings on the hose assembly or through a written certification that lists acceptable products for the hose assembly and/or restrictions provided by the owner of the hose assembly. Verification that a cargo hose assembly is appropriate for its intended use also should be incorporated into the required pretransfer procedures. The NTSB recommends that PHMSA require cargo tank motor vehicle carriers and transfer facilities to verify (1) that cargo transfer hose assemblies, whether carried on the vehicle or provided by the facility, are chemically compatible with the hazardous material to be transferred and (2) that drivers verify hoses are marked as compatible with the material to be transferred before either loading or unloading operations begin.

3.2 Inadequate Passive Emergency Discharge Requirements

Title 49 CFR 173.315(n)(2) requires that bulk transport vehicles transporting certain liquefied compressed gases, including anhydrous ammonia, be outfitted with passive emergency shutdown control equipment. The passive shutdown system serves as a means to shut off automatically the flow of product from the cargo tank motor vehicle—without the need for human intervention—within 20 seconds of an unintentional release caused by "a complete separation of a liquid delivery hose." The two types of passive shutdown systems commonly used in industry to satisfy this requirement are the Smart-Hose system found on the accident hose and a permanently mounted, computer-controlled leak detection/shutdown system. The computer-controlled leak detection/shutdown system activates in the event of pressure change, whereas the Smart-Hose system activates as a result of mechanical failure of the hose assembly that leads to the internal cable's being stretched to a predetermined length. Consequently, a computer-controlled leak detection/shutdown system can shut off the flow of product as a result of either partial or complete hose separation, whereas the Smart-Hose system can do so only in the event of complete hose separation and the tensioning of the internal cable. Both systems satisfy the current requirement to stop the flow of product from the cargo tank if there is a complete separation of the cargo hose assembly. However, in the Swansea accident, the hose assembly did not experience a complete separation, and the internal cable inside the hose assembly was unaffected. As a result, the internal cable was not stretched to the predetermined length necessary and the flow rate was insufficient to activate the flapper valves in the ends of the hose assembly, which would have cut off the flow of anhydrous ammonia in all directions.

The current regulation does not take into account a rupture of a cargo hose without complete separation. For toxic or flammable gases, such as anhydrous ammonia and LPG, the consequences of a hose rupture without separation can be just as severe as the consequences of a complete hose separation. The outcome of either scenario is the uncontrolled and free flow of the toxic or flammable gas from the cargo tank. The impact of the hose rupture in Swansea was slightly mitigated because the driver trainee was able to trip the emergency shut-off valve on the cargo tank and thereby prevent the complete release of ammonia from the cargo tank. Even so, the cloud of ammonia gas generated by the accident was sufficient to cause a passing motorist to suffer a fatal injury.

The current requirement[14] was established to prevent catastrophic human loss and property destruction that may result from the failure of a hose assembly while hazardous liquefied compressed gases are loaded into or unloaded from highway cargo tank vehicles. However, emergency discharge control should function under all—rather than only select—circumstances; that is, any hose assembly failure rather than only failures that result in complete hose separations. The NTSB concludes that given the unique hazards of toxic and flammable liquefied compressed gases, the requirements in 49 CFR 173.315(n)(2) for passive emergency discharge systems on highway cargo tanks fail to provide an acceptable level of protection against all types of cargo hose assembly ruptures.

The Swansea accident demonstrates that hose assemblies do not always separate completely if they fail, and the NTSB believes that passive shutdown systems designed to function as a result of a complete separation of a hose assembly alone should not be permitted to satisfy the emergency discharge control requirement. Therefore, the NTSB recommends that PHMSA amend the provisions of 49 CFR 173.315(n)(2) to require that passive emergency shutdown control systems for highway cargo tanks activate in the event of a partial or complete failure of a cargo hose assembly.

3.3 Hose Assembly Inspection and Testing Requirements

Title 49 CFR 180.416 establishes standards for inspecting and testing cargo hose assemblies that are installed or carried on specification MC 330 and MC 331 cargo tank motor vehicles that transport liquefied compressed gases such as anhydrous ammonia and LPG. Elements of the inspection and testing program include requirements for monthly inspections, annual leak tests, testing of new and repaired hose assemblies, and a safety check of each hose after unloading.

Under section 180.416(d), cargo tank motor vehicle carriers must visually inspect each delivery hose assembly at least once each calendar month in which the hose is in service and record the inspection date, the inspector's name, the identification number of the hose, the company name, the test date of the transfer hose assembly, and the result of the inspection (that is, pass or fail). The monthly hose inspection records are to be retained by the motor vehicle carrier until the next test of the same type is completed.

Although section 180.416(d) specifies that a hose assembly must be inspected each month it is in service, the regulation does not define what is meant by "in service." It seems logical that a hose assembly carried on a cargo tank motor vehicle has the potential to be used on any given day, and, therefore, it should be considered to be "in service." If a motor vehicle carrier does not believe that it will be used, it will not necessarily consider it to be in service and as a result, the hose assembly may not be inspected, as was the case in the Swansea accident. The

---

[14] On September 8, 1996, in Sanford, North Carolina, during delivery of propane to a bulk storage facility by an MC 331 bulk transport, more than 35,000 gallons of propane were released. The discharge hose separated from its hose coupling at the delivery end of the hose. Most of the transport's 9800 gallons of propane and more than 30,000 gallons from the storage tanks were released. If this quantity of released propane had ignited, local authorities estimated that about 125 emergency response personnel could have been injured or killed. *Federal Register* Volume 62, Number 159 (Monday, August 18, 1997). http://www.gpo.gov/fdsys/pkg/FR-1997-08-18/html/97-21865.htm (accessed July 5, 2011).

intent of this regulation presumably was to ensure that each hose assembly carried on a cargo tank motor vehicle is inspected on a monthly basis. However, the wording in the regulation seems to have created a loophole, because the term "in service" could be interpreted differently by various motor vehicle carriers. Because monthly inspections are an important preventive measure for the identification of physical deterioration, damage, and excessive wear, it is critical that these inspections are performed routinely.

Typically, if PHMSA receives an industry inquiry regarding a specific hazardous materials regulation, it publishes a formal interpretation of the regulation in question to clarify or explain the intent of the regulation. These interpretations are disseminated to the respective parties and posted on the PHMSA website. PHMSA has not published an interpretation on this subject matter. The NTSB concludes that the lack of clarity of 49 CFR 180.416(d) regarding monthly inspections of "in service" cargo hose assemblies can lead motor carriers to mistakenly defer monthly inspections of transfer hose assemblies that are carried on cargo tank vehicles but believed not to be used regularly. Therefore, the NTSB recommends that PHMSA publish and disseminate a formal interpretation of 49 CFR 180.416(d) that includes the criteria used to determine when a cargo transfer hose assembly is "in service."

3.4 Inadequate Annual Hose Assembly Leakage Test Requirements

Title 49 CFR 180.407 specifies the requirements for annual leakage tests for specification MC 330 and MC 331 cargo tanks that are used to transport liquefied compressed gases such as anhydrous ammonia and LPG. In accordance with section 180.407(h)(1), the leakage test is to include product piping "with all valves and accessories in place and operative." Section 180.407(h)(4) requires registered inspectors of MC 330 and MC 331 cargo tanks to inspect visually the delivery hose assembly for noticeable defects while the hose assembly is under the same test pressure as the tank. This paragraph further states that hose assemblies that are not permanently attached to the cargo tank motor vehicle can be inspected separately from the cargo tank motor vehicle. Title 49 CFR 180.416(e) requires that the "owner of a cargo hose assembly that is not permanently attached to a cargo tank motor vehicle must ensure that the hose assembly is annually tested" in accordance with section 180.407(h)(4).[15] Under section 180.407(h)(4), in addition to the written record of the inspection of the cargo tank motor vehicle, the registered inspector conducting the leakage test of the hose assembly must record the hose identification number, the date of the test, and the condition of the hose assembly.

The most recent annual external visual inspection and leakage test records for Werner's nine cargo tank motor vehicles that were equipped with hose assemblies were reviewed by NTSB investigators; these records were found to be incomplete. Those tests were performed by (1) L&L Repair and Testing, Inc. and (2) Boyd Service, Inc. Although each of the reports indicated that the hose assembly had been visually inspected and found to be in an acceptable condition, only three of the reports included a separate leakage test report. The test reports for the other six trailers indicated that the hose assemblies had been visually inspected and found to be in acceptable condition; however, no indication existed that any of these hose assemblies had

---

[15] These leakage tests include visual inspection of the hose assemblies while they are under leakage-test pressure (that is, 120 percent of maximum working pressure).

been leak tested, and none of the required hose identification information was included in any of the reports.

Additionally, although none of the 12 other vehicles that Werner owned at the time of the accident were equipped with hose assemblies, the test records for those cargo tank vehicles indicated that they were carrying hose assemblies that had been visually inspected and, in some cases, the test records included hose identification numbers. The inconsistencies found in the inspection and testing records strongly indicate that the registered inspectors had not been consistently conducting leakage tests on the cargo hose assemblies on the cargo tank vehicles nor were they completing annual leakage test reports as required.

The FMCSA also uncovered deficiencies in Werner's test records. During a postaccident facility review, the FMCSA cited one of the companies contracted by Werner to perform inspections because it failed to include required information on test and inspection reports and to retrain hazardous materials employees every 3 years. The FMCSA reviewed 50 test and inspection records and found that all of them were missing some of the required information. The NTSB concludes that Werner's incomplete and incorrect inspection records of cargo tank and hose testing suggest that the accident cargo hose assembly may not have been inspected and tested properly before the accident. The NTSB believes that the compliance reviews conducted by the FMCSA following the Swansea accident, and its subsequent enforcement actions, satisfied the need for an audit of Werner and its contracted registered inspectors. Therefore, the NTSB is not issuing any safety recommendations for this purpose at this time.

Notwithstanding the hose inspection deficiencies of Werner and its contracted registered inspectors, the lack of clarity of the regulation (section 180.407(h)(4)) is also a factor in this accident. After stating the requirement for registered inspectors to inspect the hose assemblies while under leakage test pressure, the regulation states, "Delivery hose assemblies not permanently attached to the cargo tank motor vehicle may be inspected separately from the cargo tank motor vehicle." Although PHMSA has not published a formal interpretation of this language, it has indicated to investigators that "inspected separately" is intended to mean that a hose assembly does not have to be physically attached to the cargo tank to be tested for leaks. As is the case with the language contained in section 180.416 regarding monthly hose inspections being completed on hose assemblies that are in service, the NTSB believes that this language could also be interpreted differently depending on the individual. One possible misinterpretation would be that a hose assembly may be tested for leaks at a time other than during the annual inspection of a cargo tank motor vehicle. PHMSA told investigators that a cargo tank motor vehicle should not pass an annual inspection without its hose assembly being leakage tested, since the hose assembly is considered to be part of the vehicle. However, the fact that registered inspectors allowed 12 of Werner's vehicles to pass annual inspection when none of them were equipped with hose assemblies indicates that industry and the regulators do not agree on the scope and procedures for leakage testing. Therefore, the NTSB concludes that the lack of clear requirements for testing cargo hose assemblies and cargo tank motor vehicles for leaks has adversely affected the accuracy of the test records. The NTSB recommends that PHMSA issue guidance to motor carriers and registered inspectors that clarifies the testing and the recordkeeping requirements of 49 CFR 180.407 for cargo hose assemblies and cargo tanks that are used to transport liquefied compressed gases to ensure that all hose assemblies are tested for leaks on an annual basis.

4. Conclusions

4.1 Findings

1. The accident hose failed because it was not chemically compatible with the anhydrous ammonia in the cargo tank that caused the chemical degradation, the loss of mechanical strength, and the ultimate failure of the cargo hose.

2. Because of the chemical incompatibilities of the hose material with anhydrous ammonia and of the couplers with liquefied petroleum gas, Werner Transportation Services, Inc.'s hose assembly was not suitable for use with either anhydrous ammonia or liquefied petroleum gas and should not have been carried on the cargo tank motor vehicle that was involved in this accident.

3. Verification of chemical compatibility of the hazardous material and the transfer hose assembly as part of pretransfer protocol and before loading or unloading begins is critical to minimizing the risk of incompatibility.

4. Given the unique hazards of toxic and flammable liquefied compressed gases, the requirements in Title 49 *Code of Federal Regulations* 173.315(n)(2) for passive emergency discharge systems on highway cargo tanks fail to provide an acceptable level of protection against all types of cargo hose assembly ruptures.

5. The lack of clarity of Title 49 *Code of Federal Regulations* 180.416(d) regarding monthly inspections of "in service" cargo hose assemblies can lead motor carriers to mistakenly defer monthly inspections of transfer hose assemblies that are carried on cargo tank vehicles but believed not to be used regularly.

6. Werner Transportation Services, Inc.'s incomplete and incorrect inspection records of cargo tank and hose testing suggest that the accident cargo hose assembly may not have been inspected and tested properly before the accident.

7. The lack of clear requirements for testing cargo hose assemblies and cargo tank motor vehicles for leaks has adversely affected the accuracy of the test records.

4.2 Probable Cause

The National Transportation Safety Board determines that the probable cause of the accident was Werner Transportation Services, Inc.'s use of a cargo hose assembly that was not chemically compatible with the anhydrous ammonia that was being unloaded. Contributing to the accident was the lack of explicit requirements by the Pipeline and Hazardous Materials Safety Administration that the motor carrier and the facility carrier verify that the cargo hose assembly is chemically compatible with the product to be transferred before transfer operations begin.

5. Recommendations

As a result of its investigation of this accident, the National Transportation Safety Board makes the following safety recommendations:

To the Federal Motor Carrier Safety Administration:

With the Pipeline and Hazardous Materials Safety Administration, jointly issue a safety advisory bulletin to inform cargo tank motor vehicle owners and operators, registered inspectors of these vehicles, and transfer facility operators about the circumstances of this accident and actions needed to prevent the occurrence of a similar accident. (H-12-1)

To the Pipeline and Hazardous Materials Safety Administration:

With the Federal Motor Carrier Safety Administration, jointly issue a safety advisory bulletin to inform cargo tank motor vehicle owners and operators, registered inspectors of these vehicles, and transfer facility operators about the circumstances of this accident and actions needed to prevent the occurrence of a similar accident. (H-12-2)

Require cargo tank motor vehicle carriers and transfer facilities to verify (1) that cargo transfer hose assemblies, whether carried on the vehicle or provided by the facility, are chemically compatible with the hazardous material to be transferred and (2) that drivers verify hoses are marked as compatible with the material to be transferred before either loading or unloading operations begin. (H-12-3)

Amend the provisions of Title 49 *Code of Federal Regulations* 173.315(n)(2) to require that passive emergency shutdown control systems for highway cargo tanks activate in the event of a partial or complete failure of a cargo hose assembly. (H-12-4)

Publish and disseminate a formal interpretation of Title 49 *Code of Federal Regulations* 180.416(d) that includes the criteria used to determine when a cargo transfer hose assembly is "in service." (H-12-5)

Issue guidance to motor carriers and registered inspectors that clarifies the testing and the recordkeeping requirements of Title 49 *Code of Federal Regulations* 180.407 for cargo hose assemblies and cargo tanks that are used to transport liquefied compressed gases to ensure that all hose assemblies are tested for leaks on an annual basis. (H-12-6)

BY THE NATIONAL TRANSPORTATION SAFETY BOARD

DEBORAH A.P. HERSMAN
Chairman

ROBERT L. SUMWALT
Member

CHRISTOPHER A. HART
Vice Chairman

MARK R. ROSEKIND
Member

EARL F. WEENER
Member

Adopted: April 12, 2012